YOUR KNOWLEDGE HAS VALUE

- We will publish your bachelor's and
  master's thesis, essays and papers

- Your own eBook and book -
  sold worldwide in all relevant shops

- Earn money with each sale

Upload your text at www.GRIN.com
and publish for free

Bibliographic information published by the German National Library:

The German National Library lists this publication in the National Bibliography; detailed bibliographic data are available on the Internet at http://dnb.dnb.de .

Imprint:

Copyright © 2015 GRIN Verlag
Print and binding: Books on Demand GmbH, Norderstedt Germany
ISBN: 9783668935372

This book at GRIN:

https://www.grin.com/document/467916

Soboka Gebisa

Assessing the total economic value of pastoralism in Ethiopia

GRIN Verlag

ASSESSING THE TOTAL ECONOMIC VALUE OF PASTORALISM IN ETHIOPIA

BY

SOBOKA GEBISA

(MSc in Agricultural Economics)

A paper submitted to the Department of Agricultural Economics for the course seminar on Agricultural Economics (AgEc 652) Dilla University, College of Agricultural and Natural Resources

June, 2015

Dilla Ethiopia

TABLE OF CONTENTS

ACKNOWLEDGMENTS

It has become the willingness of **GOD** to successfully complete this seminar paper. I, therefore, praise **GOD**.

I am grateful to thank my advisor Dr.**Neduri Suryanarayana** for their assistance in the production of this review. I would especially like to thank him for his constructive comments and feedback on all aspects of this work.

Last but very important I would like to express my heartfelt appreciation and gratitude to my Friend **Tesfaye Nenko** for the grand support and encouragement.

ACRONYMS AND ABBREVIATIONS

ASALs	Arid and Semi-Arid Lands
COPACSO	Coalition of Pastoralist Civil Society Organizations
CSA	Central Statistical Authority
EEA	Ethiopian Economic Association
FAO	Food and Agriculture Organization
GDP	Gross Domestic Product
ICPALD	IGAD Center for Pastoral Areas & Livestock Development
IIED	International Institute for Environment and Development
ILRI	International Livestock Research Institute
MOFED	Ministry of Finance and Economic Development
NABC	Netherlands-African Business Council
PFE	Pastoralist Forum Ethiopia
SNNPR	Southern Nations and Nationalities People's Region
TEV	Total Economic Value
TLU	Tropical livestock unit
USAID	United State Agency for International Development
USD	Unite State Dollar
WISP	World Initiative for Sustainable Pastoralism

ABSTRACT

Ethiopian Pastoralists make an immense contribution to the national economy despite living in some of the most inhospitable and drought-prone parts of the country. Fundamental misconceptions about the pastoral production system in Ethiopia (like in many other countries in Africa) have led to a general perception among policy makers that pastoral lands are underused and therefore should be 'developed'. Such misperceptions have subjected pastoral communities to political and economic marginalization. These economic misunderstandings have an important impact on the environmental goods and services that pastoralism provides, since they lead to low and misdirected investment, poor service provision and promotion of less sustainable alternatives to pastoralism. The aggregate results of reviews of the paper indicate that despite the widespread opinion that pastoralism is not an economically viable or rational livelihood activity; it contributes significantly to the GDP of Ethiopian economies.

This Review on the Total Economic Value of Pastoralism has two broad objectives: reviewing the Economic significance of pastoralism and the Direct and Indirect economic contribution of Pastoralism to the Ethiopian economy; by using a framework for Total Economic Valuation of Pastoralism in Ethiopia and the review contains the findings of an assessment of the Total Economic Value of pastoralism in Ethiopia, identify important knowledge gaps; Using the findings, the report discusses trends in pastoral economies and policy options that can support drylands economies more effectively.

The future of pastoralism is the subject of national and global discussions. The concerns are receiving attention from the scientific community to generate knowledge and share experiences and best practices that may offer solutions for the survival of pastoralism and the millions of people dependant on this livelihood.

Keywords: Pastoralism, Total Economic Valuation, Direct and Indirect values, Ethiopia

Part I Introduction

Pastoralism occupies a quarter of the world land area which is predominantly arid and semi-arid and supports tens of millions of pastoral households in which 60% are found in Africa. Pastoralism is an economic activity and land use system with its own distinct characteristics and it is a way of life for people who derive most of their income or sustenance from keeping domestic livestock reared in conditions where most of the feed is natural rather than cultivated or closely managed (Sandford, 1983).

According to (FAO, 2001), Pastoralism is one of the key production systems in the world and is taking place in about 25% of the globe and 66% of the entire continent of Africa. About 62 million hectares or 60% of the total land mass of Ethiopia, mostly the drier and hotter lowland parts, is inhabited by pastoralists (PFE, 2009).

Pastoralists represent some 10% of Ethiopia's population (which is about 72 million) and approximately 40% of the land area of Ethiopia is considered suitable for pastoral land use only (to be under pastoral production) (Helland, 2006). After one year the national census of 2007, show that 12 million or 20% of the total population of Ethiopia depends on pastoralism and agro-pastoralism (CSA, 2007).

The drier and hotter lowlands of the country are inhabited by pastoral populations comprising the whole of Somali region (accounting for 57% of the pastoralists in Ethiopia), the Afar region (26% of the Ethiopian pastoralists). The Borana and Karrayu pastoralists in Oromia Regional State together account for about 10% of the total pastoral communities in Ethiopia. The remaining 7% of the Ethiopian pastoralists inhabit the lowlands of the Southern, Gambella and Beni Shangul regions (Yacob A., 2000; Sandford and Habtu, 2000 as sited in Abebe D. and Solomon, 2015).

The pastoral sector makes a very significant contribution to the national income, employment, agricultural production, and food demand of people in the world. They produce 10% of the global meat used for human consumption (Roger Blench, 2001).

The contribution to national economics is commonly expressed in terms of proportion of Gross Domestic Product (GDP), which in turn, is calculated from national livestock population and production coefficients. Therefore, GDP indicators are indirect measures and in part, depend on estimates of the proportion of the national herd found in pastoral areas of the country concerned and provides an immense contribution to the national economy of Ethiopia by raising 40% of the cattle, 75% of the goats, 25% of the sheep, 20% of the equines and 100% of the camels (Yacob Arsano, 2000). The total direct economic contribution of pastoralism to the Ethiopian economy (through the production of milk, meat, skin, hides, etc.) is estimated at US$ 1.53 billion, which accounts for about 6% of the agricultural GDP per annum (Berhanu and Feyera, 2009).

However, despite their economic contribution, there has been a fundamental misunderstanding of the pastoral production system in Ethiopia (like in many other countries in Africa). Resource alienation and curtailment of mobility has made pastoral households vulnerable to frequent droughts, food insecurity and famine (Eyasu E., 2014).

Work on the economic value of pastoralism and its development potential, showing that far from being the low productivity subsistence economy it was reputed to be, pastoral livelihoods make a major contribution to GDP and in many countries to exports. This is only partially captured by national economic statistics. In Ethiopia the inclusion of more accurate calculations of the contribution of livestock, especially in pastoral production, means readjusting the agricultural GDP figures upwards by 47 percent (IGAD, 2013).

Pastoral system is not simply a mode of livestock production, rather a complex system that needs adequate and careful valuation. They are also consumption systems that support millions of mobile pastoralists globally. There is a multiple and extensive set of values associated with pastoralism: some are tangible but many are not; some can be measured but many cannot; and those that can be measured are often underestimated (Halefom Y., 2014).

1.1. Objectives

The general objective of this paper is to review of the Total Economic values (TEV) of pastoralism and to give bird's eye view on the total economic values (TEV) of pastoralism in Ethiopian Economy.

Specific objectives are to:

The Specific purpose of the present paper is to review:-

> To Review the Economic significance of pastoralism in Ethiopia
> To Review the Direct and Indirect economic contribution of Pastoralism to the Ethiopian economy.

To Review the environmental values and importance of pastoralism

Provide Conclusion and Recommendation based on the review.

1.2. Statement of the problem

❖ The danger of undervaluation of real economic values of pastoralism is that it may wrongly lead to the tendency of gradual replacement of pastoralism by other land use systems that may be economically less feasible and may impose high costs on the environment in pastoral rangeland areas.

❖ The economic contribution of the pastoral production system to the national economy is not readily measurable and is mostly understated.

1.3. Significance of the Study

This review will be significantly useful for many reasons. It will help to highlight the multiple economic values of pastoralists that inform researchers, students, donors and development partners; it my fills the knowledge gaps that exist in the area of total economic benefits of Pastoralism by exploring both its direct and indirect benefits.

1.4. Approach and Methods

1.4.1. Approach

This Review adapted to the Total Economic Value (TEV) Analysis as its overall approach for data collection and analysis. The TEV concept is now a well-established and useful framework for identifying the various values associated with such production systems as pastoralism.

1.4.2. Methodology

This reviews has relied heavily on secondary data gathered from a review of previous studies on pastoralism undertaken in Ethiopia and other African countries and existing documentation on TEV of pastoralism in Ethiopia.

PART II: TOTAL ECONOMIC VALUES OF PASTORALISM IN ETHIOPIA

2.1. The Economic Valuation Definitions and Concepts

2.1.1. Definition of Pastoralism

Pastoralism is a subsistence (economic) pattern in which people make their living by tending herds of large animals. It is most often an adaptation to semi-arid open country in which farming cannot be easily sustained (Kandagor DR., 2005).

Pastoral development is the development of a livelihood system where households make more than half of their income from livestock-related activities using some degree of mobility to access commonly managed pastures. Such livelihood systems are not necessarily based entirely on livestock – they may include some agriculture, hunting or selling charcoal for example – but livestock are the principle source of income (Kratli and Swift, 2014).

Other definitions relate to the natural resource base in the production system. According to (Dietz, 1987) pastoralism is a livestock-based or a pasture-based economy, in which livestock directly or indirectly provides more than half of the food needs of households.

2.1.2. Total Economic Value (TEV) Concepts

Total economic value is composed of use value and nonuse value. Use value represents the utility enjoyed by people who directly use the good and it require actual participation to enjoy them. On the other hand, nonuse value refers to the value that people assign to preserve the good but do not use in a commercial or other manner. Both use and non-use values can be measured using willingness to pay or willingness to accept (Tietenberg and Lynne, 2012; Hanely and Barbier, 2009; Hackett and Sharpe, 2006 sited in Halefom Y., 2014).

$$TEV = use \quad value + Non\text{-}use \quad value$$

Use values are categorized in to direct use value (DUV) and indirect use value (IUV). On the other hand, nonuse values (also called passive use value) are inherent in the good. Nonuse or passive use values consist of existence value, bequest value and option value (Perman et al., 2003).

The concept of TEV underlies that Pastoralism is a way of life that adapts to marginal environments, characterized by climatic uncertainty and low-grade resources. It has considerable economic value and latent potential in the dry lands, and is central to the livelihoods and wellbeing of millions of the worlds poor (Hatfield and Davies, 2006).

2.1.3. Total economic value of pastoralism

Two types of livestock values have been emphasized in the empirical literature. These are the direct and indirect values. Direct value involves products that can be quantified such as milk, meat, sales of livestock, savings, cultural requirement among others while the indirect value entails tangible and less tangible values like source of livelihood, social support, market assess, and food security among others (Campbell and Knowles, 2011).

It is essential to extend the concept of pastoralism economic benefits beyond the value of livestock products (milk, meat, hides) to include all "values" associated with it. These values also need to be disaggregated within pastoral systems and households and between them and the broader national economy (Ced Hesse, 2006).

Figure 1: Simplified schematic of the total economic value of pastoralism

TOTAL ECONOMIC VALU OF PASTORALISM

USE VALUES

NON-USE VALUES

Direct value – Subsistence, trade and raw material inputs; employment and skills development; and other livelihood factors

Indirect value – Inputs to tourism, agriculture and service provision; ecological and rangeland services; indigenous knowledge

Option value– Retaining future opportunities to access benefits from pastoralism

Existence and bequest value – Intrinsic benefits for global society and private values for pastoralism's preservation

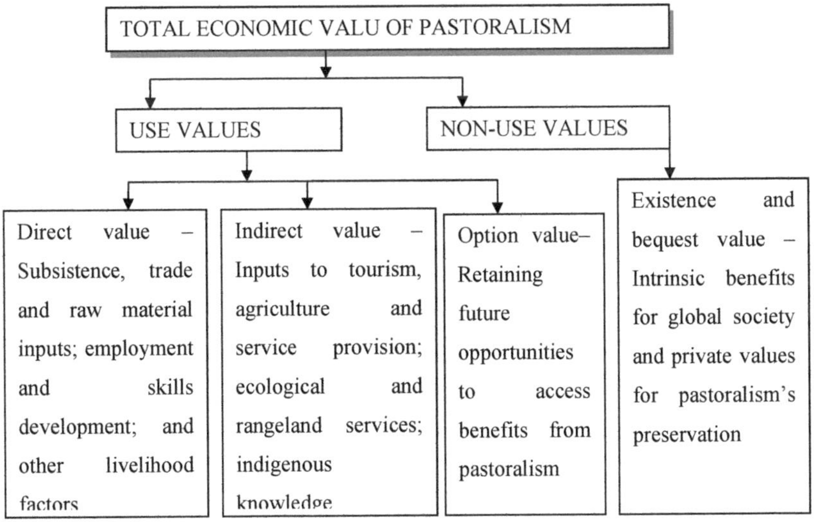

Source: (Ced Hesse and MacGregor, 2006).

Figure above presents a conceptual framework composed of four "values" with which to assess the total economic value (TEV) of pastoralism to a national economy.

2.2. The livestock economy and Economic significance of Pastoralism

2.2.1. The Pastoral Economy in Ethiopia

Pastoralism is a production system made up of people, natural resources, livestock and social relations. None can exist without land of some sort (PFE, 2010). Pastoral livestock production is more profitable than crop cultivation even where cultivation is mechanized (Little et al., 2010 as cited by Abebe M., Dejene D. and Solomon B., 2015).

Pastoralists in Ethiopia are mainly found in seven regions. The main livelihoods systems include pastoralism, farming and ex-pastoralism those who have dropped out of pastoralism and now survive on petty income-earning activities (Behnke et al., 2007).

Pastoralists constitute a minority in Ethiopia, with an estimated 12–15 million of Ethiopia's 77 million people (PFE, 2006). Livestock in pastoral regions accounts for an estimated 40% or so of the country's total livestock population (Pantuliano, S. and M. Wekesa., 2008). The Intergovernmental Authority for Development (IGAD) estimates that in 2008/09 the pastoral livestock population contributed 34.8 billion ETB (Ethiopian Birr) out of the total national livestock value of 86.5 billion ETB to the national economy.

According to the Ministry of Agriculture, Ethiopia's total livestock population has reached more than 88 million the largest in Africa and the livestock sub-sector contributes an estimated 12% to the total GDP and over 45% to the agricultural GDP (Courtenay Cabot Venton, Tenna Shitarek, Lorraine Coulter, and Olivia Dooley, 2013). In inclusion of more accurate calculations of the contribution of livestock to Ethiopia economy, especially in pastoral production, means readjusting the agricultural GDP figures upwards by 47 percent (IGAD, 2013).

Over 97% of the pastoral population lives in Somali, Oromia, Afar and Southern Region States (Table 1). It is estimated that in a country of close to 80 million people about 9.8 million people in Ethiopia are pastoralists. The data shows that out of the total 37.9 million TLU of livestock the country owns, 9.9 million TLU of livestock are found in the Pastoral areas. It shows that the share of pastoralism in the national livestock wealth is 26%.

Table 1: the Location and Distribution of pastoral areas in Ethiopia

Region	Total surface area('000km2)	Total pastoral area (km2 sq.)	Pastoral Districts	Human population ('000)	Livestock density (TLU/km2)
Afar	98.4	90.4	29	1301	7
Benishangul-Gumuz	48.4	8.4	3	40.6	1
Dire Dawa	1.2	1.2	1	108.6	32
Gambella	25.8	17.3	5	133.6	17
Oromia	353.0	152.1	34	4007.9	33
SNNP	112.3	30.4	6	219.7	23
Somali	325.1	325.1	44	4002.2	8
Total	956.1	624.8	122	9813.6	16

TLU = Tropical livestock unit, equal to 250 kg of live weight (Jahnke, 1982).

Source: (MoA and RD, 2004; (EEA, 2005).

The Economic potential of the area is immense indeed. The expensive rangelands are rich in bio-diversity. Big rivers cut across the pastoralized areas of the country. 40% of the cattle, 75% of the goats, 25% of the sheep, 20% of equine and 100% of the camels are found in the pastoralized area of the country. Almost all the National Park is situated in this area (Yacob, 2000).

2.2.2. Livestock Numbers, Products and Consumption in Ethiopia

Ethiopia has a huge livestock population. The sector also provides livelihood for 65% of the population. In different parts of the country livestock and livestock products account for 37–87 percent of the total household cash income. However, its maximum contribution to the total GDP is about 16 percent and to the agricultural GDP is around 30 percent (EEA, 2005 and Ayele et al., 2003).

In the pastoral and agro pastoral areas, livelihoods of the people entirely depend on livestock (Birhan, 2013). Significant reliance on livestock by household producers as a basic source of livelihood is one of the defining features of pastoralism. Studies indicate that an estimated minimum of 93% of the population in pastoral areas is directly dependent on livestock for subsistence (Mohammud Abdullah, 2003). The pastoral economy is also a significant source of surplus to the national economic growth. Data on pastoral livestock are significantly inaccurate, but some studies roughly indicate that the pastoral regions of Ethiopia account for 42% of the total national livestock population (Coppock, 1994; Desta and Coppock, 2002).

Ethiopia's pastoral peoples are managers of a system that produces a commodity of a high income elasticity of demand. This reflects that, if properly managed, their system can effectively benefit from growing national and global economies.[1]

Table 2: Livestock Population   Statistics (2004 -2010)

Numbers ('000)	2004	2005	2006	2007	2008	2009	2010
Cattle	38,103	38,749	40,380	43,125	47,571	49,298	50,884
Sheep	16,575	18,075	20,734	23,633	26,117	25,017	25,980
Goat	13,835	14,859	16,364	18,560	21,709	21,884	21,961
Camel	471	459	438	616	1,009	760	808
Poultry	35,656	22,605	32,222	34,199	39,564	38,128	42,053
Horse	1,447	1,518	1,569	1,655	1,776	1,787	1,995
Donkey	3,770	3,930	4,289	4,498	5,573	5,422	5,715
Mule	321	318	341	326	377	374	366
Beehives		4,546	4,020	4,884	4,800	4,600	4,598
%Annual changes							
Cattle		1.7	4.2	6.8	10.3	3.6	3.2
Sheep		9.0	14.7	14.0	10.5	-4.2	3.8
Goat		7.4	10.1	13.4	17.0	0.8	0.3
Camel		-2.7	-4.6	40.9	63.7	-24.7	6.3
Poultry		-36.6	42.5	6.1	15.7	-3.6	10.3
Horse		4.9	3.4	5.5	7.3	0.6	11.6
Donkey		4.3	9.1	4.9	23.9	-2.7	5.4

---

[1] Review of Policies and Strategies Related Pastoral Areas Development WABEKBON Development Consultants PLC July 2007.

| Mule | | -1.1 | 7.2 | -4.4 | 15.7 | -0.8 | -2.1 |
| Beehives | | -11.6 | 21.5 | -1.7 | -4.2 | 0.0 | na |

Source: Central Statistical Authority MOFED and EMA, 2011

Table 2 above displays Ethiopia's livestock statistic from 2004 to 2010. The table is divide into two subsections, the upper part showing livestock number in thousands for the corresponding categories of livestock types during the period. The cattle population has grown from 38 million in 2004 to 51million in 2010.[2] Similarly, the total number of shoats (sheep plus goats) has increased from 29 million to 47 million during the period. While the number of camels nearly doubled, the size of poultry population increased only by a relatively smaller amount population increased only smaller amount arise from 36 million to 42 million. Similarly, the increase in the number of equines have followed similar pattern of growth as that of camels (ibid).

Livestock production is fundamental in agricultural economies especially where climatic conditions and environmental location make livestock farming the most practicable option since they play multifaceted roles in the livelihoods of households who keep them (Waters-Bayer and Bayer, 1992; Randolph et al., 2007; Cecchi et al., 2010 as cited in Adesugba and Margaret, 2014).

In 2009 the official estimate of the livestock contribution to agricultural GDP was slightly more than 32 billion Ethiopian birr or $3.2 billion US dollars. In the same year, if we combine an adjusted estimate of the livestock component of agricultural GDP with the value of livestock services excluded from GDP, we arrive at a total value for livestock's contribution of 113 billion Ethiopian birr, or roughly $11.3 billion US dollars at 2009 exchange rates. In other words the total economic benefits of livestock goods and services are more than three and a half times greater than the MOFED's original estimate of the value added from livestock in 2008-09. In short, the great bulk of what Ethiopian livestock provide for the domestic economy is not identified in natural accounts as coming from livestock (ICPALD, 2013).

In the pastoral and agro pastoral areas, livelihoods of the people entirely depend on livestock (Birhan, 2013). According to (Negassa et al., 2011), in smallholder mixed farming

---

[2] Fitaweke Metaferia, Thomas Cherenet, Ayele Gelan, Fisseha Abnet, Agajie Tesfay, Jemal Abdi Ali, and Wondessen Gulilat, (2011). A Review to Improve Estimation of Livestock
Contribution to the National GDP pp 15-16.

systems, livestock provides nutritious food, additional emergency and cash income, a means of transportation, farm outputs and inputs, and fuels for consumption.

Table 3: Livelihood benefits derived from ruminant and equine livestock, 2008-09 in billion EB.

Type of Benefit	Agricultural GDP	Services not in Current GDP estimates
Value added livestock products (meat, milk, etc)[3]	MOFED: 32.232 re-estimated: 47.687[4]	
Traction power for ploughing		21.500
Benefit from financing		12.800
Benefit from self-insurance		8.600
Benefit from risk pooling/stock Sharing		3.650
Transport and haulage by Equines*		18.959*
Sub-totals	47.687	65.590
Total economic benefits	113.196	

Source: Notes: *refers to 2009-10

Total economic benefits of livestock goods and services, now estimated at more than 113 billion EB, are more than three and a half times greater than the MOFED's original estimate of the value added from livestock in 2008-09. Of the roughly 80 billion EB increase in benefits, about 15 billion EB are derived from recalculating the value of livestock products, and the remaining 65 billion come from broadening the estimation to include livestock services[5].

2.2.3. Livestock and the Ethiopian Export Economy

---

[3] Defined as the gross value of rumi nant output less intermediate costs of 408,026 EB, i.e. all of MOFED's intermediate livestock production costs except poultry feed (IGAD 2010 Table 4).
[4] Source IGAD LPI Working Paper No 02-10 (2010)
[5] Table 9 is flawed because it combines data from two years – from 2009-10 for equine power and from 2008-09 for all other outputs. This shortcomin g will be corrected in the final version of this report, which will adjust all estimates to a common ye ar based on additional data from MOFED.

For pastoralists in developing countries, livestock provides a backbone for most families and their communities as it helps provide their nutritional necessities in addition to the opportunity to sell excess livestock produce in markets to earn income (Blench, 2001). Pastoralism also contributes immensely to the gross domestic product of certain countries with contributions of between 20 to 80% in countries such Ethiopia, 40% (Hatfield et al., 2006).

2.2.3.1. Official Export of Livestock and Livestock Products

Livestock have diverse functions in the livelihood of Ethiopian farmers in the various farming system (Ehui et al., 1998; Belete et al., 2010) and serves as a source of food, traction, manure, row materials, investment, cash income, foreign exchange earnings and social and cultural identity.

The total livestock population for the country is estimated to be 43.12 million cattle, 23.63 million sheep, 18.56 million goats, 4.5 million donkeys and 0.62 million camels (CSA 2007). The livestock sector is reported to contribute 12-16% of total GDP and 30-35% of the agricultural GDP (Halderman, 2004), a figure that Behnke (2010, 2011) increases to 22% and 45%, respectively.

Similarly, Ethiopia has the largest livestock population in Africa estimated at about 70.79 million head of cattle, 28.48 million sheep, 25.91 million goats, 24.56 million donkeys, 11.39 million horses, 8.08 million mules, 8.39 million camels, 42.51 million poultry and 71.62 million beehives (CSA, 2010/2011).

Ethiopia's exports consist of live cattle, sheep, goats and camels, as well as chilled goat meat and mutton, which are mainly sourced from pastoral areas. Pre-dating the New Alliance there was an unprecedented growth in formal exports, and this trend has continued (Table 4). The main supply areas are Borana for cattle and chilled sheep and goat carcasses, and Somali Region for live camels, sheep and goats. Other supply areas include the lowlands of Bale, Southern Nations, Afar and the mid-altitude agropastoral zones of Oromia (Yacob Aklilu and Andy Catley, 2014).

Table 4: Formal live animal and meat exports from Ethiopia, 2005-2013

	Live animals		Meat	
	Number	Value (US$1,000)	Amount (tons)	Value (US$1,000)
2005/06	163,000	27,259	7,717	15,598

2006/07	234,000	36,507	7,917	18,448
2007/08	298,000	40,865	5,875	15,471
2008/09	150,000	77,350	6,400	24,480
2009/10	334,000	91,000	10,000	34,000
2010/11	472,041	148,000	16,877	63,200
2011/12	800,000	207,100	17,800	78,800
2012/13	680,000	150,000	16,500	68,000

Source: National Bank of Ethiopia (NBE) as cited in (Yacob Aklilu and Andy Catley, 2014).

Pastoral Ethiopia (population 15 million) comprises 60% of Ethiopia's land, experiences very low rainfall and frequent droughts, and has large grazing areas that hold half of the nation's livestock, which account for over 90% of meat and live animal exports (USAID, 2012)

In the highland agro-ecology of the country where crop–livestock system dominates, livestock is an essential component of the overall farming system and contributes up to 87% of the cash income of smallholders (NABC, 2010).

Livestock exports from Ethiopia have been booming for several years (albeit from a low level), and the region as a whole is ideally placed to cater to strong demand from the Middle East. The formal trade in Ethiopia earned $ 125 m in 2010 and $ 215 in 2011 while earnings from the informal / cross-border trade are estimated between US 200-300 m; 4 to 5 times the formal trade (Cately et al., 2011).

The exports generated by Ethiopian pastoralists are second to coffee in generating foreign exchange. In 2006, Ethiopia earned US$121 million from livestock and livestock products (IIED, 2010).

In Ethiopia the value of official livestock and meat exports has fluctuated widely over the decades, while official exports of hides, skins and leather have been both more stable and more valuable. For example, in the twenty-one year period from 1984 to 2004, hides and skins provided on average 90% of official livestock sector exports, livestock provided 6% and meat 4%. For a time in the 1990s, hides, skins and leather were Ethiopia's second largest export earner after coffee. The current situation is depicted in Table 6 which gives the US dollar value and percentage export share of Ethiopia's major exports from 2002-03 to 2008-09 (ICPALD, 2013).

Table 5 Below shows that the contribution of the livestock sector (live animals, meat and hides, skins and leather products) to exports has held steady at about 11% of the national total, with declines in the value of skins, hides and leather being offset by roughly comparable increases in live animal exports.

Table 5: National Bank of Ethiopia estimates of the value in million US dollars and percentage of export share for major exports, 2002-2009

Commodity	2002-03	2003-04	2004-05	2005-06	2006-07	2007-08	2008-09
Leather, Hides and skins	52.22 10.8%	43.59 7.3%	63.73 8.0%	75.0 7.5%	89.6 7.6%	99.2 6.8%	75.3 5.2%
Meat	2.42 0.5%	7.66 1.3%	14.59 1.7%	18.5 1.9%	15.5 1.3%	20.9 1.4%	26.6 1.8%
Live animals	0.481 0.1%	1.91 0.3%	12.82 1.5%	27.6 2.8%	36.8 3.1%	40.9 2.8%	52.7 3.6%
Total	11.4%	8.9%	11.2	12.2	12.0	11.0	10.6

Source: (ICPALD, 2013).

2.2.3.2. Unofficial Export of Livestock and Livestock Products

In Ethiopian pastoralist areas it has been estimated that at least 44% of the off take of cattle, 56% of the off take sheep and 30% of the off take of camel goes to export or to illegal markets (Rodriguez, 2008).

Table 6: Value (million US dollars) and percentage of export share for major exports, with and without the cross border livestock trade – 2002-03 and 2008-09.

Commodity	2002-03 Official	2002-03 cross border included	2008-09 Official	2008-09 cross border included
Leather, hides and skins	52.22 (10.8%)	52.22 (8.9%)	75.3 (5.2%)	75.3 (4.4%)
Meat	2.42 (0.5%)	2.42 (0.4%)	26.6 (1.8%)	26.6 (1.6%)

Live animals	0.481 (0.1%)	106.48 (18.1%)	52.7 (3.6%)	302.7 (17.8%)
Total Major exports + Livestock share	482.78 (100%)	588.78 (100%)	1447.9 (100%)	1697.9 (100%)
Livestock/products share	11.4%	19.66%	10.6	23.8%

Source: (ICPALD, 2013).

Table 6: above takes two widely accepted estimates for the value of informal cross border livestock trade (US $106 million in 2002-03) and (US $250 million in 2008-09) and adds these estimates to official figures for the relevant years.

Including the cross border trade, live animals were the second most important national export by value in 2002-03, following coffee, and the third most important export in 2008-09, following coffee and oilseeds. The revised total value of livestock and their products now stands at about 20% of all national exports, up from 11% according to official calculations (ibid).

During 1976, the Ethiopian government reported the estimated Birr 50-250 million illegal exports of cattle, sheep, goats and camels in its World Bank development loan application. The estimates of the concerned ministries indicated the exports of 1.1 million animals excluding camels valued at US$ 120 million in 1983. The World Bank's 1987 report estimated the unofficial exports of 225,000 cattle and 750,000 sheep and/or goats and 100,000 camels. The basis of these estimates is not very clear (Ayele et. el, 2003).

2.3. Total Economic Values of Pastoralism in Ethiopia

2.3.1. Valuation in Ethiopian Context

Following the framework provided in the project design as discussed in section 2.4 and 2.5 attempt has been made to assess/value the economic values of Pastoralism in the Ethiopian context.

Table 7: Conceptual framework for assessing the value estimates for Pastoralism in Ethiopia

Direct values		Indirect values	
Measured	Unmeasured	Measured	Unmeasured

Livestock sales (off-take) value	Employment	Input to agriculture: (Draft power, and manure)	Ecological and rangeland Services
Milk sales	Animal husbandry and rangelands management skill	Inputs to tourism	Agricultural services – including supply of breeding stock
Off-take values of hides and skin		Some dry lands products (incense and gums)	Socio-cultural values: Indigenous knowledge
Subsistence value of milk for pastoralists			

Source: (WISP, 2006 and (SOS Sahel, 2008).

2.3.2. Approaches and assumptions adopted in valuation

A very careful and the necessary effort were made to base the estimations on the available secondary data, relevant literature, previous studies, and experts' guesses. An extensive review of and reference to various reports and studies made on the pastoralism and range lands became useful in taking assumptions and deriving the necessary coefficients as regards the share of pastoralism in the national livestock economy.

Following the principle of Total Economic Valuation the approach of Benefit Transfer has been employed to infer/use some coefficients from the existing studies in similar environments or contexts and adapting them to the context under investigation (Hatfield and Davies, 2006).

Table 8: Summary of valuation approaches and techniques (IUCN 2005)

Approach	Technique	Nature of value	Advantages	Disadvantages
Benefit Transfer	Use of other empirical studies	All direct, indirect and non-use values	Small data requirements Cost-effective	inapplicability of existing studies

Source: (Rodriguez, L, 2008)

Ethiopia has a large livestock sector. According to the data series of the Ministry of Finance and Economic Development of Ethiopia (MOFED), the real GDP for the year 2005/6 was

93.5 billion Birr (10.7 billion USD), out of which the share of livestock sector is 9% or 8.3 billion Birr (954 million USD). Agriculture, Hunting and Forestry together take share of 43%.as stated in ((SOS Sahel Ethiopia, 2008).

Table 9: Summary baseline data and estimates on pastoralism Categorization of pastoralism

Factor	Number
Summary data:	
Land in country ('000 km^2)	1.13 million km^2
Drylands in country	(60 % of national landmass); (678,000 km^2)
Pastoral land	(51% national landmass); (624800 km^2)
Pastoral population	9,813,600 (MOARD, 2004)
Density pastoral land	(16/km^2)
Number of cows	8,310,427
Pastoral owned cows on drylands	(20%)
GDP	93.55 billion Birr (2005/06), MOFED
Contribution of livestock sub-sector	10.29 billion Birr
Contribution of the Agriculture sector	40.27 billion Birr
Pastoral production:	
Milk: national and significance of pastoralism; per capita consumption and significance of Pastoralism	(2.7 billion liters); %; 14.5 liters per annum per capita; % 41% from pastoralism, (camel 114.8 million liters)
Meat: national and significance of pastoralism; per capita consumption and significance of Pastoralism	586,000 Tones; %; 10.25kg per annum per capita; 150000 MT (pastoral/lowlands); (Camel, 3120 MT)
Hides – national and significance of pastoralism	15,400 MT; 390 million Birr Pastoralism
Export of live animals – national and significance of pastoralism	41565 (official); 90% pastoralism, cross boarder /unofficial, 1,476,000 animals, value = 1,242,000,000 Birr (Pastoralism 100%)
Export earnings – national and significance of Pastoralism	4,069.90 million Birr; livestock 10.7%, pastoralism (unofficial export, via Somali and Moyale (1.58 billion Birr)
Other indirect values:	
Tourism – GDP contribution and other data	6.13 billion Birr; only 2 national parks are out of lowlands. 50% lowlands (3.06 billion Birr)
Agriculture – GDP contribution and	40.3 billion Birr (45% of national); pastoral land

other data (draft power)	contribution 20% of 14 million working age cattle for ploughing (2800000 cattle); estimated value = 1.15 billion Birr
Other – gum and incense production	National (3,457,000 Birr); pastoralism/dry lands (100%)
Fisheries: national; Pastoralism/lowlands	National production 95,000 MT; lowlands 47,000 MT

Note: The time official exchange rate is 1 USD = 9 Eth. Birr.

Source: (SOS Sahel Ethiopia, 2008).

It is necessary to reassess the national livestock and human population database in the pastoral areas upon which this valuation is based. The national livestock size in TLU is estimated to be 39,700,000 out of which the pastoral areas share 9,996,800 (25%). The national statistics on the livestock size in Ethiopia is very loose, it is more so when it comes to the pastoral livestock population. Different reports provide `different figures. According to the (MOA,2000) out of the total estimated livestock population of the country, the pastoral areas constitute approximately 30% of the cattle, 52% of the sheep, 45% of the goats, and 100% of the camels.

(Bruck, 2004) reported that a livestock population estimates obtained from the pastoral areas raises these figures to 49% of the cattle, 47.5% of the sheep, 51.5% of the goats, 100% of the camels and 12.9 % of the equines. The estimate of the pastoralist population provided by the (MOARD, 2004) is 9,813,600 while other reports commonly indicate 12-15 million (Alemayheu, 2006). Here a multiplier of 1.22 to 1.53 is implied. Referring to (Bruck, 2004), the share of Pastoralism in total cattle and shoats populations is larger i.e. 49.50%, 51.50%, 47.50%, respectively, for cattle, goats and sheep, instead of the lower value reported by others, as 30% for cattle, 45% goats and 52% for sheep. It shows that pastoral livestock size (cattle, camels, and shoats) is underestimated by a factor of 1.33 if the percentage shares are applied to the figures of (FAOSTAT, 2004).

2.3.3. Assessing the Direct Economic Values of Pastoralism

Direct values of pastoralism consist of measurable products and outputs such as livestock sales, meat, milk, hair and hides. Employment, transport, knowledge and skills are less easily measured values included in this category. Livestock in the pastoral areas are the major source of food (milk and meat) and income, as well as a source of employment (Rodriguez, L, 2008).

If equal contribution from the highlands and low lands (simply taking the size of the livestock) is assumed, the value of pastoralism for the National GDP is 2.59 billion Birr (0.28 billion USD). According to some estimates the country losses/ unaccounted for value of unofficial cross-border trade amounting to 138 million USD (1.24 billion Birr) per year. If the value of cross-border trade is added to the official GDP, then the contribution of pastoralism will rise to 3.8 billion Birr (0.43 billion USD) per annum. This shows the share of pastoralism in the Agricultural GDP stands at 10%. Similarly, this figure shows the contribution of Pastoralism to the total national GDP of 4.1% (ibid).

According to some other reports (Haramata, 2006) in Ethiopia economists suggest that pastoral production accounts for almost 20% of Ethiopia's GDP. Similarly, (Hatfield and Davies, 2006) reported a rough estimate that shows the share of pastoralism in agricultural GDP of 35% in Ethiopia. It is difficult to explain why so much variation between the different authors. It may be possible that there could be differences in the components of the Pastoralism are valued, the national data used in estimation, and underlying assumptions.

2.3.3.1. Animal sales and consumption

The value of animal sales is consistently one of the highest direct values of pastoralism across the country studies. The case studies, using average market prices for their respective countries, estimated that for the year 2006 the sales of livestock produced in pastoral systems reached values ranging between USD 192 million for Kyrgyzstan, a country with a total pastoralist herd size of about 1.8 million TLU, to over USD 2300 million for Spain, a country with a pastoralist herd size of about 8.6 million TLU. It is important to highlight that for some countries with similar herd sizes like Mali (8.4 million T LU) or Ethiopia (9.8 million TLU), the sales values are significant lower than the Spanish figures (USD 428.5 million for Mali and USD 364 million for Ethiopia). In Ethiopia the unofficial trade and illegal cross-border sales were estimated at USD 138 million per year, i.e. about 38% of all the country animal sales (Rodriguez, 2008).

Only about 60% of animal sales in Ethiopia go recorded, with the transactions that remain invisible estimated to represent a value of 138 million USD per year (ibid).

2.3.3.2. Milk production and valuation

There are four major milk production systems in Ethiopia. These are pastoral and agro-pastoral, smallholder crop–livestock mixed system, urban and peri-urban, and intensive dairy farming (Azage and Alemu 1998). Pastoralist and smallholder farmers produce 98% of the country's total milk production (CSA 2008). Total milk production in 2005 was estimated at 1.5 million tonnes which is equivalent to USD 398.9 million (FAOSTAT 2007).

In Ethiopia the milk produced by pastoralists represents about 65% of the national milk production, but the estimated total value of the milk produced in pastoralists systems in official statistics amounts to only USD284 million. The figure is greatly underestimated owing to the high proportion of milk that is consumed within the household and therefore not captured by markets or statistics: at least 77% of the total milk produced in pastoral areas (Rodriguez, 2008).

A sample survey of 237 pastoral households interviewed in Yabello, Borana zone and Karayu pastoral communities. The sample survey data shows that 77% of the milk produced is consumed by the households while 16% is sold. Similarly, they consume 91% of the produced butter and sell 7% of it (ibid). The challenge of measuring milk production remains an obstacle for appropriate valuation and has led to great underinvestment in this important pastoral subsector (Davies and Hatfield, 2007).

Table 10: Milk utilization in pastoral areas of Ethiopia

Pastoral Areas	Amount home Consumed (%)	Amount Sold (%)	Others including in-kind wage (%)	Home consumed Butter (%)
Afar	83.34	10.12	6.53	74.6
Somaile	67	28		50
Borana	59	13	21.21	
South Omo	84	7.08	9.16	
Average	73.34	14.55	12.3	
This study (sample survey)	77	16		

Source: CSA data sample enumeration, 2001/02: reported 2003

According to Sample Enumeration data (2001/02) (CSA, 2003) In Ethiopia About 15% of the milk production is sold to the market, valued at 384 million Birr (37 million USD) per annum. Close to 120 million liters of produced milk (12%) is sued for other local sues including by-products production processing and payments as in kind wage. This amount is valued at 287 million Birr (31 million USD). The total subsistence value of milk production is valued` at 2.26 billion Birr per annum (246 million USD). Including the other uses at local level and sales, the total value of milk production from pastoral areas reaches 2.65 billion Birr per annum (294 million USD) (Hatfield, R and Davies, J., 2006).

Pastoral areas in Ethiopia are generally known as rangelands and cover about 0.7 million square km (two-thirds of total area). In 2010, a total of 2,940 million liters of milk were produced from about 9.6 million cows at national level (FAO, 2011).

Table 11: Number of Cows for milk production* Ethiopia 2006-2011

Year	2006	2007	2008	2009	2010	2011
Quantity	4,422,000	5,153,000	7,600,000	9,919,368	9,627,747	10,676,783

Source: FAOSTAT, 2013

*refers to a cow that is primarily kept for milk and has been milked previously or was being milked at the time of enumeration or has never been milked before but expected to be milked in the future or pregnant at the time of enumeration (NABC, 2013).

Table 12: Producer price of fresh cow milk (USD per ton), Ethiopia 2006-2011

Year	2006	2007	2008	2009	2010	2011
USD per ton	303.5	413.5	505.2	498.3	362.3	270.90

Source: FAOSTAT, 2013

In five reference years, 2005–2009, export values increased from about 73 000 USD to 123 000 USD, while import values increased from about 5.6 million USD to 10.3 million USD during the same period (NABC, 2013).

2.3.3.3. Hides and skins sales and consumption

Hides and skins exports have historically been a major export category for Ethiopia, but there is no recorded data on hides and skins trade from pastoral areas (Aklilu, 2009 cited in Peter D. Little ,Roy Behnke ,John McPeak and Getachew G., 2010:36).

In terms of the overall monetary, Ethiopian hides and skins account for 85% of the country's livestock exports. The value of the export of skins and hides to the national economy was estimated in USD 600 million, however the estimated value of hides and skins for pastoralists amounts only USD 43 million. This amount is much less than expected considering that pastoralist herds represent over 30% of the total national animal population[6]

2.3.3.4. Leather export

The livestock sector in Ethiopia accounts for 16% of the national and 27-30% of the agricultural GDP. 13% of the country's export earnings are due to leather and live animals exportation (MOARD, 2007). Ethiopia's leather export market, dominated by the pastoralist sector, provides 12% of total trade (EPA, 2003).

2.3.3.5. Meat or the sales of livestock

Productivity of meat production in arid and semi-arid pastoral areas (Jahnke, 1982 as cited in WISP, 2008) is 16.3 kg /TLU of mixed herd. Given the estimated pastoral heard size (9.8 million TLU) in Ethiopia, the total meat production is estimated at 162,948 MT. This figure closely matches with the study report by (Diao et al., 2004) which indicates that the lowland zones/pastoral areas produce 150,000 MT of meat (bovine meat, mutton and other meats all together). This amounts to 29% of the 580,000 MT of meat production at the national level.

Pastoral areas not only meet most of the domestic meat demand but are also the main suppliers of livestock for export, generating about US$50 million per annum for Ethiopia (Yakob and Catley, 2010). In Ethiopia, pastoralists and agro pastoralists produce 14% of cattle meat, 12% of goat meat and 7% of sheep/lamb meat, and leather exports provide 12% of national export earnings.

Table 13: Estimated total and proportional meat production from pastoral systems in Ethiopia (millions)

Cattle meat		Goat meat		Sheep meat	
Pastoral/agro-Pastoral production	% of national production	Pastoral/agro-Pastoral production	% of national production	Pastoral/agro-Pastoral production	% of national production

---

| 39,254,569 | 14% | 3,006,528 | 12% | 1,839,164 | 7% |

Source: Ced Hesse & James MacGregor (2009)[7]

The focus on livestock and meat productivity in policies and development programmers' overlooks the crucial difference that pastoral systems (as opposed to the meat-focused ranching models often promoted in their place), as well as supplying the markets with low-cost meat, also support a wide range of other goods and services – starting from a substantial milk economy (often controlled by women and therefore critical to their livelihood and position in the society) with an important subsistence value (Davies, 2007).

Most farming pastoralists in the dry lands make money from their livestock and use the lower and more unpredictable returns from farming in order to save money (or reduce the intake on their 'capital' stock). Producers may also use livestock to save money, as a way of securing highly valued food (meat and milk), while avoiding penalizing market costs (Behnke, 2006).

2.3.4. Assessing the Indirect Economic Value of Pastoralism

Indirect values associated with pastoralism include tangibles such as inputs into agriculture (manure, traction, and transport) and complementary products including Gum Arabic, Honey, medicinal plants, wildlife and tourism. The contribution of pastoralism to agriculture in terms of draft power, tourism and wild life, manure for land fertilization, animal traction and incense and gums to a limited extent are captured.

2.3.4.1. Value of Draft power for farming

A report by the Ethiopian Economic Association (EEA, 2005 cited in WISP, 2008) indicates that about 59% of the 10 million crop farmers in the country keep 14 million working age cattle used for ploughing and threshing, while the rest (42%) do not have oxen. Also Pastoralism contributes to 20% of this draft power. In terms of animals this equals to 2,800,000 cattle (oxen).

---

[7] Arid waste? Reassessing the value of dryland pastoralism.

Livestock contribution to farm income in the highland is 24% and this increases to 30% when draft power and manure utilities are included. The average real farm household income during the year 2004/5 was 9545 Birr. Considering a total of 10 million farming households, the contribution share of draft power will be 5.7 billion Birr; and 572,700, 000 Birr (63 million USD) is due to contribution of draft power from Pastoralism.

Table 14: Estimated input of Pastoralism to agriculture in draft power

Input to Agriculture:	Quantity
Real per capita income/consumption expenditure, highland (Birr), 2004/05	1909
Average household size (No)	5
Total household income (Birr)	9545
Pastoralism contributes to highland draft power (%)	20
Draft power contributes to household income (%)	3
Total number of crop farmers (No	10,000,000
Total value-add by draft power to rural household (Birr)	5727000000
Total value by the Pastoralism to draft power value-add (Birr)	572700000
Total value by the Pastoralism to draft power value-add (USD)	US$ 63,633,333

Source: (WISP, 2008)

2.3.4.2. Tourism industry

The total area under the national parks and games which are found in the low land areas is estimated at 466,640 hectares (Bruck, 2004).

There are cultural and historical tourist sites in the northern parts of the country which also attract lots of tourist. Taking the GDP contribution of the tourism sector, the value share of pastoral areas will be 3.06 billion Birr (340 million USD) per annum. The tourism industry is reported to contribute 4.5 % to the national GDP.

Table 15: The estimated contribution of Pastoralism to Ethiopian Tourism industry

Item	Quantity

Tourism (GDP value), value (%)	4.5
Value of tourism industry (Birr)	6,133,600,000
Assumed share of the lowland pastoral areas (%)[20]	50
Estimated share of lowland pastoral areas (Birr	3066800000

Source: (Rodriguez, L, 2008)

2.3.4.3. Contribution of pastoralism to agriculture

2.3.4.3.1. Animal traction:

Ethiopia was estimated that about 42% of farmers do not have oxen for traction, although most of them have other livestock species. The study suggests that about 20% of the animals used for traction in Ethiopia are provided by pastoralists.

Using this methodology, an Ethiopian study showed that the use of traction could increase the farmers' income by 6%, from a baseline of about USD1060 per year. The aggregated figures suggest that considering a total of 10 million farming households and 20% of the traction power provided by pastoralists' animals, the contribution of traction power from pastoralism to the national economy could be about USD155.6 million (Behnke, R. 2006).

2.3.4.3.2. The Value of Manure Production

The use of manure could increase the national production of wheat by 1.29 million tons. Using market prices to quantify the increment in monetary terms, the final figure suggests that the value of manure from pastoral systems to Ethiopian agriculture should be about USD 160 million per year (Rodriguez, L., 2008).

The national account of 2005/6 reported by (MOFED) shows that that the value of manure/animal dungs production was 902,182,000 Birr. Taking a share of Pastoralism in the national livestock size (265) and assuming no difference in the amount and quality of animal manure produced in the low land and highlands gives share of Pastoralism that equals to 237,966,570 Birr.

2.3.4.4. Production of incense and gums

The pastoral communities protect these trees partly as a feed for browsers and also since these trees generate income for households by producing gums and incense. Although there is a large potential of benefiting from incense and gums resources in the lowland areas, the country generates a limited amount of income, 3,457,000 Birr per annum, from this resource. A 100% of this can be attributed to the drylands (WISP, 2006).

In General The total sum of measured indirect values of Pastoralism including values form draft power, manure production, tourism and incense and gum greater than 4.12 billion Birr (over 485 million (USD). This amount is almost as much as the direct values of Pastoralism

Part III. Concluding and Recommendations

3.1. Conclusions

The major purpose of this review is to shed light on the total economic values (TEV) of Pastoralism in Ethiopia and add to the knowledge base towards sustainable development and support for Pastoralism. The review is interested in the application of the concept of Total Economic value to explore, identify and value the direct and indirect, values of Pastoralism with the aim of laying a strong base for lobbying and advocacy for sustainable development of Pastoralism. Given this objective, however, what is achieved in this seminar paper is the assessment of the direct and indirect values of Pastoralism in Ethiopia, while the aspects of option and existence values could not be addressed.

The review of the literature clearly shows that Pastoralism has a huge value for the national economy and the pastoralists themselves. Such a huge economic value is undermined because of many reasons including the lack of appropriately disaggregated database that reflects the system itself. Above all, the marginalization for Pastoralism in terms of access to basic social and development services has hindered the realization of the huge economic value.

Based on reviewed literature data, the total measurable direct and indirect values of the total economic value of Pastoralism in Ethiopia is in Billions of Birr. This is a significant value that Pastoralism is contributing to the national economy beyond supporting the life of pastoralists themselves.

Pastoralism is a serious economic contributor to Ethiopia's economy and could be an even more serious contributor to domestic trade and export earnings, given the high value of the subsistence economy. Indirect values of pastoralism are also poorly understood, but pastoralists have a major role to play in service provision to a wide national and international clientele.

3.2. Recommendations

Recognize the livestock and plants in pastoral areas are a crucial element of Ethiopia's genetic resources and thus, should be protected on that basis; support research and extension that responds to needs and interests of pastoralists, and which draws on their extensive indigenous knowledge.

The official statistics must capture all the economic values associated with pastoralism in GDP database

Decision makers should compare Economic benefit between pastoralism and other more intensive uses of the land.

Pastoralism is the only means of producing food without replacing the natural vegetation so appreciate and value pastoralism in Ethiopia will makes Ethiopia the future indigenous supplier of Livestock product in World.

Developing Methodology that should be better to measure the current and potential contribution of pastoralist production to Ethiopian livestock product outputs,

Pastoralism Economics must needed as school of thought in Ethiopian as another African country since pastoralism contributes in billion of Birr in to Ethiopian Economy.

Pastoralists will need better access to credit and savings institutions to improve animal husbandry.

Because of lack of appreciation, many pastoralist systems have collapsed, Modernize the image of pastoralism and create incentives.

References

Abebe, M., Dejene, D. and Solomon, B. (2015). An approach to Securing Pastoral Land Rights in Ethiopia Tetra Tech ARD, Ethiopia Paper prepared for presentation at the"2015 World Bank Conference on Land and Poverty" The World Bank-Washington D.,

Aklilu, Y. and Catley, A., (2010). Livestock Exports from Pastoralist Areas: An Analysis of Benefits by Wealth Group and Policy Implications. IGAD LPI Working Paper No. 01 - 10.

Aklilu, Y. and Catley, A., (2014). Pastoral Livestock Trade and Growth in Ethiopia: for the Future Agricultures Consortium Policy Brief 72 p 2.

Azage, T. and Alemu, G/Wold. (1998).Prospects for Peri-Urban Dairy Development in Ethiopia In: Proceedings of 5th national conference of Ethiopian Society of Animal Production (ESAP), 15–17 May 1997, Addis Ababa, Ethiopia. p 248.

Behnke, R. (2006). Review of the literature on Pastoral Economics and Marketing: The Horn of Africa and Southern Africa Report prepared for the World Initiative for sustainable Pastoralism, IUCN EARO. Odessa Centre Ltd., UK. 2006

Behnke, R., (2010). The Contribution of Livestock to the economies of IGAD member states. IGAD LIP Working Paper No 02 -10.

Behnke, R. and Osman, H.M. (2011). The Contribution of Livestock to the Sudanese Economy, IGAD LPI Working Paper No 01-12, Djibouti.

Belachew, H. (2004). Livestock Marketing and Pastoralism. In Yonas (Ed) Pastoralist Forum Ethiopia, Ethiopia, Addis Ababa.

Bruk, Y. (2003). Food Security Situation in the Pastoral Areas of Ethiopia A research report.

Ced Hesse and MacGregor, J. (2006). Pastoralism: drylands' invisible asset? Developing the framework for assessing the value of pastoralism in East Africa. International Institute for Environment and Development (IIED). Issue No.142 pp.5-24.

Coppock DL. (1994). The Borana plateau of southern Ethiopia: Synthesis of pastoral research, development and change, 1980–91. ILCA (International Livestock Centre for Africa), Addis Ababa, Ethiopia. Pp. 23-34

CSA. (2003). Report on Farm Management Practices, Livestock and Farm implements. Agricultural Sample Enumeration of 2001/2002, Ethiopia, Addis Ababa.

CSA. (2007). Ethiopian population and housing census.

Desta, S. & Coppock L. (2004). Pastoralism under pressure: tracking system change in Southern Ethiopia. Human Ecology 32 (4): pp 465–486.

Davies, J. (2007). Total Economic Valuation of Kenyan Pastoralism, WISP-IUCN, Nairobi Pp.2-4.

Dietz, T. (1987). Pastoralism and Dire Straits: Survival Strategies and External Interventions in a Semi- arid at Kenya/Uganda border: West Pokot, 1900 – 1986.Nederlance Geografische Studies 49.

EEA. (2004/05). Transformation of the Ethiopian Agriculture: Potentials, Constraints and Suggested Intervention Measures. Report on the Ethiopian Economy. Volume IV 2004/05. Addis Ababa.

Halefom, Y. (2014). Households Willingness to Pay for Camel Milk in Aba'ala Woreda, Afar Regional State Mekelle, Ethiopia pp. 9-13.

Hatfield, R. and Davies, J. (2006). Global Review of the Economics of Pastoralism World Initiative for Sustainable Pastoralism (WISP), IUCN, Nairobi..

Helland, J. (1999). Land Alienation in Borana: some land tenure issues in a pastoral context in Ethiopia Eastern African Social Science Research Review, Vol.XV, No.2.

Helland, J. (2006) Pastoral Land Tenure in Ethiopia Régime foncier pastoral en Ethiopie Chr. Michelsen Institute, Bergen, Norway pp. 5-6.

Hesse and McGregor, (2006).Pastoralism: drylands' invisible asset? Developing a framework for assessing the value of Pastoralism in East Africa IIED issue paper No. 142

Hussen K., Tegegne A., Kurtu M. Y. and Gebremedhin B., (2008): Traditional cow and

Camel milk production and marketing in agropastoral and mixed crop–livestock Systems: The case of Mieso District, Oromia Regional State, Ethiopia. IPMS (Improving Productivity and Market Success) of Ethiopian Farmers Project Working Paper No. 13.

IGAD, (2013) Member States Discuss the Action Plan 2014-2018 for the Regional Migration Policy Framework, IGAD.

IIED, (2013). Pastoral Total Economic Evaluation. Workshop Report, 29th January to 1st February 2013, Addis Ababa, Ethiopia, International Institute for Environment and Development, Edinburgh.

ILRI (International Livestock Research Institute), (2014). WISP. A global perspective on the total economic value of pastoralism: Global synthesis report based on six countries valuations. Nairobi, Kenya. 56 pp. 11-24

Independent Office of Evaluation of IFAD and the Office of Evaluation of FAO Concept note (2014). Joint Evaluation Synthesis Report on FAO's and IFAD's engagement in pastoral development.

Kandagor, DR. (2005). Rethinking Pastoralism and African Development: A Case Study of the Horn of Africa. Egerton University, Njoro-Kenya

Kerven and Behnke, R. (2011). Launching Pastoralism as An Open access Journal with Springer Open. Pastoralism: Research, Policy and Practice.

Ketema, H. Tsehay, R. (1995). Dairy production systems in Ethiopia. In: Proceedings of a workshop entitled:

Krätli, S. (2014). If Not Counted Does Not Count? A programmatic reflection on methodology options and gaps in Total Economic Valuation studies   of pastoral systems IIED Issue Paper. IIED, London. Pp. 10-16.

Pastoralist Forum Ethiopia, (2008). Proceedings of the Fourth National Conference on pastoral development in Ethiopia Millennium Development Goals and Pastoral Development: Opportunities and Challenges in the new Ethiopian Millennium UN ECA Conference Hall, Addis Ababa.

Pastoralist Forum Ethiopia, (2002).Proposed Pastoral Development Policy Recommendations

Submitted to the Ministry of Federal Affairs Ethiopia, Addis Ababa.

PFE, IIRR and DF. (2010). Pastoralism and Land: Land tenure, administration and use in pastoral areas of Ethiopia Published 2010 by PFE, IIRR and DF. pp .1-9.

Peter D. Little, Behnke, R. McPeak, J. and Getachew, G. (2010). Pastoral Economic Growth and Development Policy Assessment Future Scenarios for Pastoral Development in Ethiopia, 2010-2025 Report Number 2, Ethiopia p. 12.

Rodriguez, L. (2008). A global perspective on the total economic value of pastoralism: global synthesis report based on six country valuations Nairobi pp. 4-21.

Solomon, Z. A., Binyam, K., Bilatu, A. Ferede, A. and Gadisa, M. (2014). Smallholder cattle production systems in Metekel zone, northwest Ethiopia pp.1-3.

SOS Sahel Ethiopia, (2008). Pastoralism in Ethiopia: It's Total Economic Values and Development Challenges, IUCN-WISP, Nairobi.

SOS Sahel Ethiopia, (2008).Policy Framework for Pastoralism in Africa: Securing, Protecting and Improving the Lives, Livelihoods and Rights of Pastoralist Communities page I, 9.

Swift, J. (1988). Major Issues in Pastoral Development with Special Emphasis on Selected African Countries FAO, Rome.

WABEKBON Development Consultants PLC, (2007). Review of Development Policies and Strategies Related to Pastoral Areas in Ethiopia

Yacob, A. (2000). "Pastoralism in Ethiopia: The Issues of Vi ability", In Proceedings of the National Conference on Pastoral Development in Ethiopia, Addis Ababa: Pastoralist Forum Ethiopia. pp. 2-3.

Yakob, A. and Catley, A. (2010). Livestock exports from pastoralist areas: an analysis of benefits by wealth group and policy implications. Inter-Governmental Authority on Development Livestock Policy Initiative Working Paper No. 01-10.

Zander, K. (2006). Modelling the Value of the Borana Cattle in Ethiopia an approach to justify its conservation. Bonn, Germany: Center for Development Research.